Contents

Acknowledgements, Introduction & Dedication

Acknowledgements

Special thanks are due to Dr Matthew Pead of the Royal Veterinary College; Ben Walton of the Small Animal Teaching Hospital, Liverpool University; Richard Payne MA VetMB AdvCertVPhys CertEdVPT HonFIR VAP MRCVS; Dr Tom Lewis and Ruth Dennis of the Animal Health Trust in Newmarket; Dr Shahad Mohammed, veterinary physiotherapist at the WitsEnd Rehabilitation Centre, Leicester; Natalie Lenton of the Canine Massage Therapy Centre in Worcs; Lisa and Mark at the Suffolk Canine Country Club; Gemma White at Petsavers charity; Nick Thompson, The Holistic Vet; Catherine O'Driscoll of Canine Health Concern; Guy Blaskey of Pooch & Mutt, and Lucy Kemp of Miramar vets.

Also big thanks go to the editors of *Dogs Today* and *Dog Monthly* magazines; BBC Radio Norfolk and Eastern Daily Press (EDP) newspapers, for their help in finding case studies.

Finally, I owe a huge debt of gratitude to the marvellous owners who've got in touch to share their dogs' inspiring stories with me, and to so kindly and generously send me so many wonderful photos of their firmest friends enjoying life to the max.

Introduction

Just like people, dogs are living longer, which, of course, is a good thing. The only downside is the older they get, the more likely they are to develop arthritis, although any dog – whatever their age, breed or size – can fall victim to the disease.

An estimated one million dogs have some form of arthritis, so it's clearly a common problem which can strike at an early age. And, of course, this figure could be even higher as some dogs are never diagnosed with the disease, slipping through the net to soldier on regardless.

Arthritis is a painful condition, and one of the main causes of chronic pain in dogs. It can badly affect their quality of life and ability to enjoy the fun activities they so love. It's also a complicated disease, for dogs as well as humans, with many forms and different causes. Sometimes there's no known trigger.

While osteoarthritis, or degenerative joint disease, is the most common type in dogs, they can also develop auto-immune versions similar to rheumatoid arthritis in humans. Other forms include infective arthritis, perhaps as a result of a bite, an allergy or a reaction to certain drugs or vaccines.

None of us likes to see our dogs suffer, so it's important to get a firm diagnosis as soon as possible to enable appropriate treatment to begin, which will allow your dog to resume his or her place at the centre of family life. The signs of arthritis are usually clear to spot if you're prepared to look, even if you can't hear what your dog's trying to tell you.

If only dogs could speak, they'd be able to say that their knees are a little stiff this morning because they ran around just a bit too much yesterday, or their hips hurt when they try to climb the stairs. Or they don't want to be petted today, thank you all the same, because they're hurting.

If your dog's favouring the sofa over a walk in the park, this could be a sign that all's not well. And being so immobile means that it's easy for your dog to gain weight, subjecting joints to extra pressure. By the same token, if your dog's off his food for a while, or losing weight without a reason, coupled with any personality changes, it's time to take your dog to the vet for a proper check-up.

Until canines can talk (surely it's only a matter of time?), you have to be your dog's advocate – and best friend – and ensure he gets the help he needs and deserves. Your dog will thank you for it, from the tips of his paws to the tops of her ears!

While an arthritis diagnosis can be daunting, there's a lot that can be done to help relieve pain and stiffness, using an holistic approach of conventional drugs, complementary therapies, a healthy diet, weight loss, and proper exercise. Surgical procedures can be carried out if your dog's arthritis is severe enough to warrant this.

It is this book's aim to help you to help your dog, with expert veterinary advice; heart-warming case histories; treatment options; tips for coping, and recommendations on diet and supplements. There are also chapters covering physiotherapy; complementary therapies; hydrotherapy, and where to go for further information.

It's always worth sharing your stories with other owners by joining a local club, or by taking part in one of the increasing number of online forums for specific breeds, which discuss everything from concerns over health to the latest pet aids and health supplements. New social networking sites, purely for dog owners, are springing up, too. Some nice vets post comments and advice online so you could well pick up an extra tip or two, free of charge, which can't be bad.

Finally, stay positive: chances are your dog certainly will, as our case histories attest. Below are some of the happy hounds we'll be meeting along the way, living their lives to the full.

Dedication
To all the dogs who haven't let a little thing like arthritis stop them in their tracks

Chapter 1

The different types, and signs to look for

An estimated 20 per cent of our canine friends are living with osteoarthritis. The condition accounts for the majority of vet visits, and is one of the most common causes of pain a vet's likely to treat.

Arthritis isn't a straightforward disease, and just why it occurs – and to certain dogs – isn't always clear.

But as a general rule, a mixture of nature (genes) and nurture (a trigger such as an injury or infection) is to blame.

So what does arthritis mean for your dog, and what are the signs to look out for?

What is canine arthritis?

The word 'arthritis' derives from the Greek words for joint (arthro) and (itis) inflammation.

The term arthritis is now generally used as a catch-all for a group of conditions affecting not just stiff and inflamed joints, but also the muscles, tendons, cartilage, bones and ligaments supporting the joints.

Arthritis follows a different pattern in dogs than it does in people, and is usually a secondary problem to a joint defect like hip or elbow dysplasia, or an injury such as a tear to a cranial cruciate ligament in the knee (stifle).

Degenerative joint disease (DJD)

This is the most common type of canine arthritis, and is rather like osteoarthritis (OA) in humans. But whereas humans tend to develop OA as they age, dogs with degenerative joint disease can be relatively young. To understand DJD, it's perhaps helpful to look at how a dog's joints work.

SYNOVIAL JOINT

The correct name for the type of joint in which arthritis can develop is a synovial joint, where two or more bones

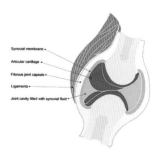

Typical synovial joint.

 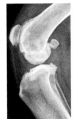

X-rays showing normal knee (left), and arthritic knee. (Courtesy Animal Health Trust)

are in contact with, and move against, each other: for example, in the knee joint.

The joint is enclosed by a tough capsule lined with a layer of synovial membrane, which secretes fluid to lubricate and nourish the joint tissue.

LIGAMENTS
These thickened parts of the capsule help to control movement. The knee has tough ligaments on both sides which allow the joint to bend and straighten, but prevent it from bending sideways and rupture. There are also strong ligaments running through the middle of the knee called cruciate (cross-over) ligaments which prevent the shin bone (tibia) from sliding too far forward or backward under the thigh bone (femur). Ligament tears are common in active dogs, and puppies who want to run before they can walk.

CARTILAGE
Hyaline cartilage is one of the key tissues in a healthy joint, acting as a shock absorber and providing a smooth surface between the bones. Cartilage

Dulcie (left) was clearly finding walking difficult, and would struggle to keep up with Daisy.

Rosie, an English Springer Spaniel and Pets as Therapy dog, owned by Richard

"Rosie is 15, and has been suffering from arthritis for a couple of years. I first noticed her arthritis when her headlong dashes through the undergrowth to spring game slowed down.

"I've always had her coat stripped in the warmer months – she loves getting into water and up to her knees in mud – and after having her done, I noticed a definite thickening of the joints, especially in her front legs.

"She also became unsteadier on her back legs on rough ground, and would fall frequently when turning, but she would still walk for miles, tail wagging all the time. As the year wore on she became more reluctant to do the long walks, and would just stop and look sorrowful. And, by golly, she can do that very well! I only had to say 'do you want to go back?' and she would immediately turn and trot off back to the car.

"After her stroke the situation became a lot worse. It has affected her eyesight and she can get rather lost when out in open spaces without a reference point to guide her.

"I've been giving her a glucosamine and chondroitin tablet each day for over a year now. They're the same ones that I take. I spoke to the vet at the time and he said to carry on. She has anti-inflammatory tablets, too, and they're helping; along with a couple of short walks every day. She prefers not to walk on rough ground as she used to, though, so we avoid that.

"But her tail still wags all the time; she can't wait to go out in the car, and eats her meals like a snowplough!"

Sammy, a nine-year-old male Greyhound who's been re-homed by the RSPCA centre in Norwich

Animal Care Manager Claire says: "Sammy was signed over to us by his elderly owner when he couldn't afford veterinary treatment for him. He arrived very underweight and in poor condition.

"Sammy initially received treatment for severe dental disease, overgrown claws, and a skin problem thought to be caused by fleas. He was taken to the vet for a health check, where it was noticed that he had some arthritis in his hips; the right was worse than the left.

"He was put on a 27kg dose of Meloxivet once a day and 1 Synamine Forte Plus msm capsule three times a day.

"The vet has advised that the Meloxivet can be reduced in the long-term if Sammy improves, but that the joint supplement should be ongoing.

"Sammy is an easy, laid-back chap; happy to be handled, and friendly towards new people and other dogs, though he doesn't like cats! He thoroughly enjoys being fussed over and is typical of his age and type – chasing small furries and cats.

"He likes a walk but is happy to curl up on the sofa for the rest of the day. He's now been re-homed with one of our volunteer dog walkers who fell in love with him."

consists of an arrangement of collagen fibres filled with hyaluronic acid and chondroitin sulphate, among other components. It's very slippery – twenty times more so than ice.

How DJD develops

For various reasons, which aren't always clear, changes occur in the structure of a joint, causing it to become inflamed, and it's this inflammation that starts the disease process and causes all the damage.

As the joint beomes stiff and swollen, the cartilage is overloaded, and can't do its job of cushioning the ends of the bones, which then grate against each other.

Moving around becomes difficult and painful for your dog as a result.

Signs to look out for

Dogs are stoical by nature, and are used to hiding pain, but their distress is probably clear to see. It's not normal for your dog to hold her paw off the floor, for example, so this could be a sign that all is not well.

Other symptoms to watch for include:

- Stiffness before he gets going
- Reluctance to walk, climb stairs, or play
- Doesn't want to jump onto the sofa for cuddles

Sophie, a Golden Retriever, lived to the grand age of 17. Her owner, Catherine, believes Sophie developed vaccine-induced arthritis in her paws from the age of six. By the time she was 15, Sophie's arthritic paws were deformed.

- Limping
- Slower than usual on walks
- Isn't eating
- Lameness (with or without signs of pain)
- Is a little withdrawn (although dogs with DJD can still seem as happy as a clam)

Inflammatory arthritis

This kind of arthritis is rare in dogs: a vet could see and treat just one case of inflammatory arthritis for every 200 or so dogs with DJD.

Inflammatory arthritis is usually described as either infectious, or immune-mediated.

INFECTIOUS ARTHRITIS

This can develop as the result of a puncture wound – from a nail or a thorn – to your dog's skin; perhaps a bite, or even an infection such as E Coli. An immune response could follow.

IMMUNE-MEDIATED ARTHRITIS

This is similar to auto-immune conditions such as rheumatoid arthritis or lupus in humans, where the immune system malfunctions, and the body begins attacking its own cells. A dog's entire system – including eyes and kidneys – can be affected, as well as a number of joints rather than just one (polyarthritis). Once contracted, it's generally a lifelong condition.

Just why immune-mediated arthritis occurs isn't always clear. It could be because of a dog's genetic background; another underlying problem, such as a tumour; a virus, such as the canine distemper virus, or following vaccinations.

When there's no known cause, and no underlying disease, immune-mediated arthritis is described as 'idiopathic.'

Signs to watch for with inflammatory arthritis:

- Discomfort or pain in more than one joint
- Shifting weight from one leg to the other
- Walking awkwardly
- The inflamed joints might be warm to the touch or swollen
- Lethargy
- Fever

As the condition generally affects the whole body, your dog is very likely to appear miserable and out of sorts.

Act quickly

It's important to get a proper diagnosis for arthritis, and begin treatment as soon as possible, particularly with inflammatory forms of the disease to avoid irreversible joint damage.

Not the end of your dog's world

While it might be distressing to learn that your dog has arthritis, the good news is that a lot can be done to relieve her pain and discomfort so she can still live life to the full, running around and chasing squirrels – or whatever takes her fancy!

Chapter 2

Joints affected

A dog's joints can be subjected to stress and strain at the best of times, especially if he is carrying around extra weight. Dogs predisposed to joint abnormalities because of their genetic make-up could well develop degenerative joint disease, which injury and lifestyle can only aggravate.

Common areas for joint problems in dogs, particularly certain breeds (see chapter 5), are the elbows, shoulders, knees and hips.

Osteochondrosis & Osteochondritis Dissecans (OCD)

Osteochondrosis is a disease in which bone and cartilage in a dog's joint doesn't develop properly, leaving a loose piece of cartilage flapping around that is susceptible to injury.

Osteochondritis dissecans, or OCD, can develop as a result, causing pain, lameness, and restricted movement.

A number of joints can be affected, including the elbow, shoulder, knee (stifle), and hock (ankle), and osteoarthritis can develop over time.

The disease usually affects younger dogs and larger breeds, and can be due to genetic background, over-feeding as a puppy, and/or rapid growth.

Generally, treatment centres on painkillers or anti-inflammatories, although surgery is an option in more severe cases, and might reduce the risk of arthritis later on.

Hip Dysplasia (HD)

The hip joints attach the body to the hind legs, and carry the bulk of a dog's weight.

When the hips fail to develop in the normal way, this is known as hip dysplasia (abnormal growth). It's a common problem, particularly in certain breeds such as Labradors.

BALL AND SOCKET JOINT

The hips have ball and socket joints – one for each leg – which look rather like a caravan tow bar. These should fit snugly, allowing a dog to walk normally.

In cases of hip dysplasia the joint is often too shallow, and the ball too small and out of shape to fit as it should, allowing the muscles, ligaments, and tendons around the joints to become loose, with the result that the hip joint dislocates and becomes unstable.

HD normally occurs during a young dog's growing stages and might affect

Fern, a ten-year-old Rhodesian Ridgeback, owned by Eileen & Geoff

"Fern came to us aged five via the Ridgeback rescue. We were told she was a healthy girl, and her one and only ailment had been 'growing pains.'

"One year on from her re-homing Fern started to develop a slight limp in her front legs, and X-rays revealed what our vet described as 'textbook osteochondrosis.' It turned out the 'growing pains' were actually the start of cartilage detaching from the end of her long bones. She'd been fed food which was too high-energy (our vet described it as 'rocket fuel for dogs') as a pup, and grew too quickly. Fern now has arthritis in her shoulders and hocks.

"She's on a combination of Previcox, four-weekly Cartrophen injections, daily Seraquin supplements, a bio-flow collar, a daily lunch of sardines, pilchards or mackerel, a little less weight, and an absolute minimum of an hour's exercise a day (she will take two in her stride). Her squirrel-chasing ability is undiminished: if anything, keeping her weight down has enhanced it!"

Harvey, a 3-year-old Fox Red Labrador, owned by Hannah

"Although both of Harvey's parents had very good hip scores, when he was around 7 months old, I noticed him 'bunny hopping,' sitting down to eat, pivoting on his bottom, and occasionally carrying one of his back legs for the odd step. I monitored him for a few weeks, and reduced the amount of exercise he was having (although it wasn't much anyway, due to his age), but there was no improvement. I took him to the vet who diagnosed hip dysplasia.

"They put him on glucosamine and anti-inflammatories for a week, after which he had a follow-up consultation. This went on for about six weeks; I saw the vet a few times during this period but he was reluctant to do anything further due to Harvey's age.

"At nine months Harvey was X-rayed again, and hip dysplasia was confirmed in both hips. He had very badly-formed balls and sockets, so was referred to a specialist, who saw Harvey at 12 months old (as he was still growing he didn't want to see him beforehand), who took more X-rays and did tests.

"At 13 months, Harvey had a complete hip replacement in his right leg. All was okay – he was walking on it, etc – but he had to have crate-rest for eight weeks. Two days after coming home he wouldn't weight-bear on it.

"At this point he had to go back for more X-rays, and the vet found the new hip ball had slipped by 2mm. They wanted to leave it for a few days to see whether it would be necessary to re-operate, or leave it and hope it would be okay once it had properly settled down, although he would probably have a slight limp for life. We left it, and thankfully all was okay – and no limp!

"Two years on from the operation, people are amazed to learn that Harvey had a hip replacement as he charges around like a greyhound! He has about 45 minutes' walk every morning, and then potters about fields every evening. The vet gave him the nickname Scooby-doo because of how lively he was after coming round from the op. On walks he must cover at least double the distance of my other Lab as he never stops running. He loves to jump off banks and into ponds at great speed – often my heart is in my mouth, but no damage has been done!"

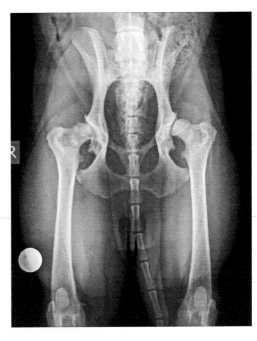

Harvey's X-rays showing his hip before (right) and after his operation.

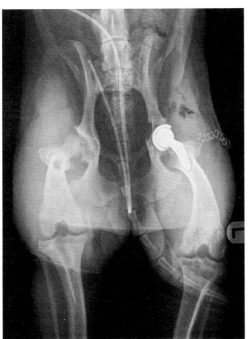

both hips. Dysplasia can also develop in the elbow.

Possible causes

Why some dogs develop HD is unclear, but the genes a puppy has inherited from her parents, combined with rapid growth, over-feeding, and/or over-exercising at critical stages of a puppy's development could be causes. Slight injury to the hip at important stages of development might have something to do with it, too.

Symptoms

In general, dogs with HD tend to limp, and might refuse to flex their hind legs because to do so is painful. (If you try to lift your dog's leg yourself, chances are he will cry in pain, and go off you for a while.) It's even been known for some dogs to hop instead of run in an effort to protect the painful part of their legs.

Symptoms also tend to vary according to age: puppies of less than three months can appear a little clumsy, and then up to around 18 months might have difficulty running or climbing stairs. The hip will usually become arthritic when a dog reaches middle age.

With young dogs the treatment will usually centre around painkillers and anti-inflammatories, although surgery might be the only answer for some.

Hip score

The British Veterinary Association and the Kennel Club have been running a 'hip score' scheme, covering around 120 breeds, for the past 40 years. Dogs of around 12 months of age are X-rayed, and each hip is given a score: the lower the score, the better.

While the scheme allows breeders (and potential owners) to be aware of hip abnormalities, particularly with Labradors, it isn't a guarantee that your

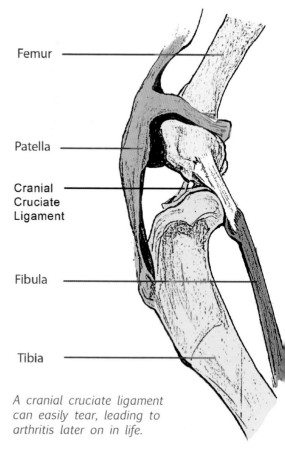

Femur

Patella

Cranial Cruciate Ligament

Fibula

Tibia

A cranial cruciate ligament can easily tear, leading to arthritis later on in life.

dog won't experience problems at some point in her life.

Injuries

A tear to a cranial cruciate (cross-over) ligament is very common. The ligament seems to gradually weaken in dogs, particularly in larger breeds, although any dog can suffer a tear, almost at any time, and whatever they are doing.

'Weekend warriors' – dogs who are over-exercised at weekends with not much activity during the rest of the week – are rather prone to this type of injury. A good clue to whether your dog has damaged this ligament is if he sits with the affected knee straightened out

Emily, an 11-year-old English Cocker Spaniel, owned by Lyndsey

"We noticed at the end of last year that Emily was slowing down when going upstairs. Then, one morning, she yelped if you touched her back, and didn't want to move. X-rays revealed she had lumbosacral spondylosis.

"Emily was prescribed Tramadol and Nutraquin, and, in just a week, the difference was amazing. We've reduced the painkiller dosage now, started her on acupuncture and hydrotherapy, and she's bouncing around again.

"I know we can't cure her Spondylosis but, hopefully, we can control it."

in front, rather than tucked-up. Such an injury usually leads to stifle (knee) arthritis over time. Painkillers and rest usually help, and surgery to stabilise the knee could delay the development of osteoarthritis.

KNOCK-ON EFFECTS

Research has shown that about a third of dogs with this injury in one knee will develop the same problem in the other knee, perhaps because of shifting weight onto the 'good' leg. Again, surgery to stabilise the knee could delay osteoarthritis developing.

Lumbosacral Spondylosis

This condition affects the lower part of a dog's spine, and usually develops in older dogs, sometimes leading to spinal osteoarthritis. While the exact cause is unclear, genes and trauma can be contributory reasons.

Symptoms often appear in the early stages of the disease, and include back pain, stiffness, and loss of balance. Painkillers are often prescribed, and your vet might also suggest alternative therapies.

Visit Hubble and Hattie on the web: www.hubbleandhattie.com & www.hubbleandhattieblogspot.com · Details
of all books · Special offers · Newsletter · New book news **twitter**

18

Chapter 3

A diagnosis: what now?

Lameness, limping, and stiffness could be sure signs that your dog has arthritis. Rather than just speculate about what might be at the root of the problem, however, it's important that you take your dog to your vet for a proper diagnosis. The earlier the confirmation of arthritis, the sooner treatment can begin, and your dog can start enjoying life again.

A thorough check over

Typically, a vet will conduct a full physical examination, looking for any swelling in your dog's joints, and any thickening around knees and elbows.

How your dog is standing – perhaps shifting weight forward if he has hind leg arthritis – is also relevant and important. And how quickly and easily your dog goes from a standing position to sitting down (some dogs with problems have a tendency to go into reverse before flopping down in a bit of a heap) will be considered, too.

FURTHER TESTS

X-rays might be ordered to reveal if there's any joint damage, although this might not show up in the early stages.

MRI scans and ultrasound tests usually provide a clearer picture of what's happening.

Some synovial fluid will probably be taken from the affected joint to help differentiate between auto-immune, infectious, and osteoarthritis.

A blood test might also be arranged by your vet to see if your dog has the rheumatoid factor (RF), although not all dogs with rheumatoid arthritis test positive to this. Blood tests will also indicate the kind of medication, should it be needed, that your dog could tolerate.

Your role

As you're the closest person to your dog, you'll probably be asked about her daily routine, and how it's changed, so be prepared to give your vet a clear picture of life at home.

For example, do you have other dogs, and do they play together? Where does your dog sleep – on the floor, on your bed, in a kennel outside – and for how long each day? What does your dog eat, and has her appetite changed recently? Or perhaps she has become a bit of a couch potato lately, which could be a sign that exercising is painful so she'd rather not bother.

Is your dog happy to play with other dogs? Your vet will take this into account when assessing her.

If there's anything you don't understand, say so, and ask your vet for an explanation. If treatment's suggested and you're worried about your dog having to take strong drugs, ask about the side effects, and possible alternatives.

Ruling out other problems

Getting a firm diagnosis of arthritis will also rule out any other health issues, such as a slipped disc, muscle strain, an infection, or another underlying condition like cancer. And it's better to get your dog checked out as soon as possible so action can be taken quickly; particularly in cases of inflammatory disease, as delayed treatment can lead to permanent joint damage.

Tips for coping

After a definite diagnosis of arthritis you might, understandably, be worried, but there's no need to panic – help is at hand. Treatment possibilities include physiotherapy or hydrotherapy and supplements, which might help manage some of the symptoms, alongside conventional treatment (see separate chapters).

Arthritis clinics

A popular course of action for many owners is to make use of the arthritis clinics provided by a number of veterinary practices. They're usually run by one of the practice's veterinary nurses once a month, and are free (although any treatment will incur expense). The clinics offer you a chance to chat to the nurse about how any

Stella the Rottweiler, owned by Lisa

"Stella's six years old, and was diagnosed with slight hip dysplasia at ten months. She showed slight lameness on a foreleg on and off over a couple of months, but I put this down to muscle strain as it only showed briefly.

"During Christmas Eve 2011 she became very lame on her left foreleg, and, as she was still quite lame two days later, I arranged a visit to the vet. An X-ray revealed that Stella has osteoarthritis in both her elbows, although her lameness had vanished.

"As the lameness resolved so quickly, the vet thought that this could have been due to a strain, so it was pure coincidence that the osteoarthritis was found.

"Stella tends not to show signs of pain, and it's worrying to think that the osteo had been developing without any obvious symptoms, and was only discovered by pure luck. Thankfully, there's still time to take action to maintain the joints as much as possible, without having to resort to surgery.

"The vet's advice was to stay off anti-inflammatories and painkilling drugs due to her young age and lack of lameness, and suggested putting her on a supplement such as Omega 6 instead."

Max, a six-year-old Westie, owned by Jean

"About 18 months ago Max was showing signs of arthritis in his right hind leg, which the vet thought was due to his putting out his hip by jumping up and down the stairs.

"The vet suggested I took Max along to the arthritis clinic so that the nurse could check him regularly, and I could chat about any worries I had.

"Max was very happy to go there – he got lots of attention. I used to take him once a month and now it's every two months.

"When his patella (knee cap) was

popping in and out of its groove, Lucy, the nurse, said she thought it would need investigating and sorted out the further checks. (Thankfully Max didn't need an operation and it popped back in.)

"People laugh when I say we're off to the arthritis clinic, but it helps so much. Max is such a lovely, affectionate chap, and I want him to get all the help he needs.

"He's my guardian angel."

drugs your dog might be taking are working, and to discuss whether the amount can be reduced and when (although older dogs in pain can't come off so quickly or easily). There's usually an opportunity to talk about any changes to symptoms and how your dog, and you, are coping. The clinics can help both before and after a diagnosis.

What else could help?

Ensure your dog has some peace and quiet away from noisy family activities; give him a gentle massage as this can aid circulation and increase flexibility. Apply a cold compress (wrapped in a dry tea towel) to the affected area for ten minutes or so a couple of times a day for three days, followed by a heat pack twice a day for ten minutes over a two week period.

Mr Schnorby's bed helps support his joints.

Cosy coats

Although dogs might not really need coats – they have fur, after all – it's probably a good idea to put one on your dog in cold weather, so that his muscles are more likely to relax and be less prone to injury.

There are so many dog raincoats, jackets and coats to choose from to keep our canine friends warm and dry, that you can take your pick. (Maybe your dog will want to have a say in this?)

You might also feel a little better about going for a daily walk, which is so important for arthritic dogs, knowing that your dog's protected from the worst of the weather: you can really feel the heat underneath a coat when you take it off your dog after a walk.

A good bed

It's worth considering investing in an orthopaedic bed for your dog: perhaps one with a memory foam filling and waterproof cover. Some are made of faux fur or suede for extra comfort; others are heated.

The pain of hip dysplasia will be worse for your dog if she has to sleep on the floor or a hard surface, so avoid this at all costs. Whichever bed you choose, make sure it's the right size for your dog's weight.

Ramps

If your dog's jumping days are over, and he's finding it difficult to get in and out of the car or onto the sofa, why not consider getting a ramp or steps which can be lightweight, are often foldable, and will reduce impact on his joints?

Elevated feeders/water bowls

Bending over to eat and drink is not comfortable for dogs with muscle/joint problems. Older dogs often eat less, anyway, and if eating and drinking are painful or uncomfortable, they are likely to eat and drink even less, easily becoming dehydrated.

Lilliput's harness could help prevent shoulder, back, and neck problems later on in life.

Harness

A rear or front support lifting harness could prove very useful if your dog has weak back legs, and needs some help with walking or being lifted. Always check with your vet first which model is the most appropriate for your dog's weight.

Dog strollers

Hubble was always happy to go in her stroller as it meant she could still go on walks with Hattie and Immie.

Similar to those for toddlers, a stroller could help if your dog tires easily. Although some dogs take a while to get used to this new method of transportation, others take to it with enthusiasm. One lady told me her 15-year-old Cairn took some convincing, but now loves riding in it, as her other dogs trot alongside.

Diesel, a 7-year-old Labrador Retriever, owned by Joanna

"A couple of years ago Diesel had a bit of a fight with a Border Collie, and began to limp.

"After a while I realised we needed to determine what the problem was, so I took him to the arthritis clinic, where he was checked out. X-rays revealed he had arthritis.

"The nurse said he was a little overweight at 36 kilos, but now he's down to 28 kilos and isn't allowed to exceed this (he's weighed regularly at the clinic). I take him to the clinic every month – or more often if I'm really concerned.

"Diesel takes Rimadyl anti-inflammatories, and painkillers on prescription.

"I haven't got insurance but I don't care how much it costs to get him fixed, and the vet hasn't overcharged me; being able to go to the arthritis clinic for advice without having to pay for it helps!"

Chapter 4

Treatment

While there's no magic bullet for arthritis, there is a range of drug treatments that will effectively manage the inflammation and pain that your dog is experiencing, and help improve her quality of life.

Drugs can form part of an holistic approach to controlling your dog's symptoms, alongside regular exercise, weight loss, physiotherapy, and complementary treatments.

Hopefully, in no time at all your dog will recover the spring in her step: tail in the air and with a smile on her lovely face once again.

Conventional drugs

Understandably, you might not be too happy about your dog taking strong drugs, but there's usually no effective alternative – particularly in the early stages of the condition. Once the inflammation is under control, the level of medication can be reduced gradually over time (although some dogs don't cope too well if medication is reduced, or removed altogether).

Some of the medication your vet might prescribe could include:

- Analgesics (painkillers) such as Previcox or Tramadol
- NSAIDs (non-steroidal anti-inflammatory drugs) like Metacam or Rimadyl. These are the newer Cox 2 inhibitors designed to be gentler on your dog's stomach
- Steroids such as Prednisolone (Prednisone) also reduce inflammation, usually very effectively and quickly. A steroid injection can provide almost immediate relief for a tender, swollen joint

Inflammatory arthritis

Dogs with inflammatory arthritis will need more treatment for longer: either steroids, or disease-modifying cytotoxic drugs which kill the cells that are dividing quickly. Treatment starts with high doses, which are then reduced to the lowest possible levels.

Sometimes there's complete remission, but your dog might need low levels of the medication (perhaps a dose every other day) to control the symptoms.

Side effects

No drug is without side effects, and sometimes these can be severe, particularly if used over the long term.

Penny, a 15-year-old Norfolk/Yorkshire Terrier cross, owned by Jane

"I got Penny from Battersea Dog's Home when she was 14 weeks old. She's a wonderful dog who never leaves my side, and she's coped with her osteoarthritis very well.

"She was diagnosed at nine years of age, and referred for an operation on her front leg. The plan was to put a plate between two bones to stop them rubbing together and causing her pain. Unfortunately, the operation couldn't be completed as the smallest available plate was still too large, and her bones would have shattered. Instead, the vet and I decided the best for Penny would be pain-killing drugs, supplements, and hydrotherapy.

"She began limping again recently, and was getting slower and slower when walking, so the vet suggested trying Cartrophen. She had one injection a week for four weeks, and now she's on a monthly injection. This really does seem to be helping.

"I'm happy we're trying everything possible. At times I forget she even has arthritis and so does she. I'm sure she'll be around for a long time yet."

Simba, a ten-year-old Staffordshire Bull Terrier, owned by Sue

"I've had Simba since she was a six-week-old puppy, and she's a wonderful companion – everyone who meets her wants to take her home with them as she is so well-behaved and affectionate. She has the heart of a lion and is my loyal defender, following me around wherever I go, sleeping outside my bedroom door each night.

"Simba had a double operation about two years ago to repair a ruptured cruciate ligament, and had a prosthetic piece put in her hind leg. The vet warned me she might be left with a limp, and was likely to develop arthritis in that leg. Sure enough, following the operation, we found that while Simba (who's as game a dog as you'll ever find) still ran after the ball with the other dogs, she was virtually incapacitated the following morning.

"Anyway, she's recovered well from the operation, and with a little help from the magnetic collar that she wears, is actively running around the hills again."

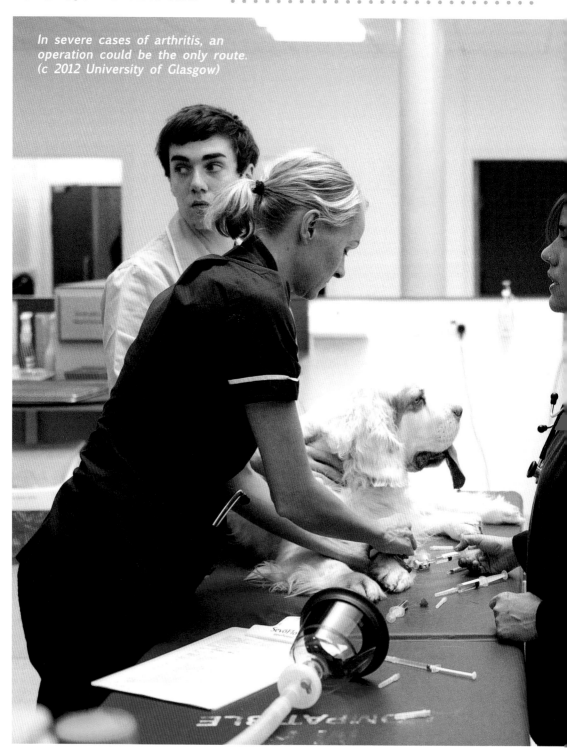

In severe cases of arthritis, an operation could be the only route.
(c 2012 University of Glasgow)

Most vets are of the opinion, though, that dogs can generally tolerate the side effects of strong drugs better than people can. Regular health checks and blood tests for liver and kidney function should minimise the chances of your dog experiencing an adverse reaction to a particular drug.

Always raise any concerns you might have with your vet, and discuss whether there's an alternative drug which will keep the symptoms under control but with fewer or less severe side effects.

Every dog is different, and some drugs might work better than others for an individual, so it could be a case of trial and error at the start of treatment.

Natural medication

Natural drugs might cause fewer side effects, but they could also be less effective; particularly in severe cases of arthritis which need more powerful pain relief.

Your vet might prescribe Cartrophen and Pentosan, which contain naturally-occurring plant substances that can help repair damaged cartilage. These are normally given as injections once a week for four weeks, and the effects can last for up to six months.

Hyaluronic acid injections into the joint, which lubricate the cartilage, reduce pain, and improve flexibility, might also be suggested. As the benefits are only temporary, injections will need to be on-going.

Human drugs

Dogs metabolise NSAIDs differently from humans, so don't be tempted to give your dog something like Ibuprofen, for example, which can be toxic for them, or anything else lying around in your bathroom cabinet for that matter.

Visits to the vet

If your dog has osteoarthritis, you'll probably see your vet on average around twice a year after diagnosis (or more often if symptoms worsen). Dogs with inflammatory arthritis are more difficult to treat, and will need more veterinary attention – possibly for life.

If your dog's arthritis is outside your vet's realm of expertise, or your dog is not responding well to the current treatment plan, he might be referred to a pain specialist, a surgeon who specialises in joint replacement, or perhaps a vet trained in acupuncture.

Centres of excellence

In the UK we're very lucky in that we have a number of leading veterinary centres around the country for referrals.

The Royal Veterinary College, the longest-standing veterinary college in the UK, can provide consultations and surgery for both osteoarthritis and inflammatory conditions at its Hertfordshire campus. The Small Animal Teaching Hospital at the University of Liverpool specialises in osteoarthritis, and has a centre dedicated to understanding and improving canine mobility.

The University of Glasgow has a Small Animal Hospital department, and a leading specialist in inflammatory arthritis, Professor Bennett, among its staff.

Fitzpatrick Referrals in Surrey, and Dick White Referrals in Newmarket are specialist veterinary centres for joint replacements.

When surgery is the only option

If your dog's joint is severely damaged, or if the pain is intense, surgery might be the only solution. While it's not without risk, particularly for overweight dogs, the right surgery can reduce pain and improve movement, and provide the nearest thing to a cure.

There are different kinds of surgery, ranging from a total hip replacement to more minor procedures.

Arthroscopy

This is the least invasive type of surgery, involving the surgeon making small cuts around the shoulder, elbow and knee joints through which to remove cartilage damage.

Fusing joints

This procedure uses metal implants with which to fuse damaged joints, particularly in dogs with inflammatory forms of the disease (repairs using bone grafts are being considered currently).

Artificial joints

Replacing a damaged joint – such as one affected by hip or elbow dysplasia – is another option. These artificial joints are similar to those used for human hip and knee replacement (although the ones for elbows are very different), and are made of a combination of cobalt chrome, steel or titanium, and polyurethane. A prosthetic (artificial) hip will normally outlive your dog and should give him a new lease of life.

Cost considerations

Prescription drugs can be expensive, and a possible alternative is to buy the drugs online. If you choose to do this, you must ensure that you use a reputable online site that insists on a valid prescription, which your vet will be able to supply (although will probably make a charge, which is reasonable).

It's also worth considering taking out a good pet insurance plan to cover most (or all) of the costs, should your dog need surgery, expensive treatments, or lots of tests: read the small print of any plan you are considering to check what is – and isn't – covered.

Stem cell therapy

Stem cell therapy (SCT) introduces new cells into damaged or dysfunctional tissue to treat a range of diseases, including canine arthritis. In the US and Australia, SCT has already become an established treatment for canine osteoarthritis, and is available at Dick White Referrals in the UK.

An artificial knee (stifle) joint. (Courtesy Fitzpatrick Referrals)

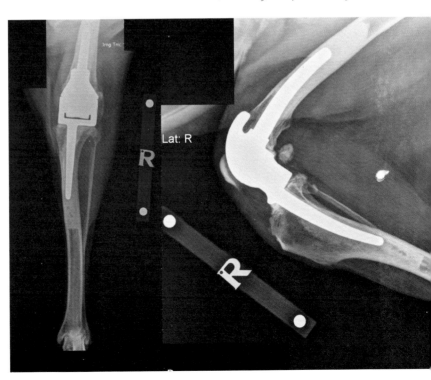

Dick White's own dog, Meg, a 15-month-old working Labrador Retriever

"Meg's elbow lameness was so pronounced – and not even controlled with continual NSAID therapy – that her career as a working dog was looking very uncertain. Meg was one of the first (if not *the* first) dogs in the UK to receive SCT (stem cell therapy) for her elbow osteoarthritis.

"Four weeks after SCT treatment, however, Meg's lameness has resolved, and she is already back working – this time without a limp and no medication.

"We have now treated four dogs, all with lameness due to osteoarthritic elbows, and all are showing considerable improvement in their lameness.

"Reports from other countries suggest that, typically, an improvement in lameness is seen within ten days, and the therapeutic effect of the SCT continues for many months, if not longer, to the extent that, in many cases, only a single treatment is required."

Chapter 5

Breeds affected

There's a misconception that osteoarthritis only affects larger breeds, such as Labradors and German Shepherds, and older dogs. Yet any dog, be she cross-bred, pure bred, or mongrel, can develop the disease, even when still young.

Add to this inherited joint problems, injury, or an infection, and the risk of contracting arthritis becomes ever more likely.

Osteoarthritis & certain breeds

Middle- and larger-sized pure breds – particularly those weighing over 25kg – tend to develop osteoarthritis (or degenerative joint disease) from about six years of age.

This category of canine is more at risk because, generally, the dogs carry more weight, and are usually the most active. An estimated 75 per cent of Golden Retrievers, Labrador Retrievers, German Shepherds, Great Danes, and Bull Mastiffs develop this form of arthritis.

The bigger the better?

We consistently choose larger dogs, such as those mentioned above, to be our faithful friends. Clearly, we like our hounds to be big, with round, friendly faces, and nice, chunky tails that wag ecstatically, even if we've only been out of sight for a few minutes.

As we favour these generously-sized breeds – and there are simply more of them around in the dog population – it stands to reason that these dogs are likely to make up the largest proportion of arthritic cases and visits to the vet, which may have given rise to the fallacy that only big breeds get arthritis.

Getting older

Although dogs can begin developing osteoarthritis when they've barely left puppyhood, the older a dog gets, whatever her breed, the more likely she is to suffer because of joint wear and tear.

And if high-impact exercise has been part of a dog's early life, and he is still darting around chasing balls, and generally doing what dogs do, in his golden years the joints will be showing their age.

In this instance, it's important to make adjustments and allowances for an older dog so that he can still enjoy his favourite games, remaining happy and free from pain.

Trudy, a German Shepherd Dog with a dash of Border Collie, owned by Jan

"I got Trudy when she was 15 months old. Around two years ago, she started getting stiff, and looked to be in pain after I'd taken her a few times around the football pitch near where I live. It turned out she had problems with her hips and her spine (DRM Spondylosis). I gave her green-lipped mussels and 'You move,' and these seemed to help.

"Towards the end she clearly wasn't into walking and would get all ploddy and slow, yet when she got home she was like a different girl, and would run up and down, playing with her ball. I taught her to lie down so she didn't have to stand for too long, and put her joints under pressure. She would roll the ball back to me with her nose – sometimes as often as 20 times at a go.

"Playing this way kept her going: the mental stimulation stopped her becoming depressed. She reached her eleventh birthday; she was my one and only Trudypops."

Bella, 7, a Springer Spaniel owned by Dick and Annie

"Bella was the runt of the litter, rescued as a puppy by a drugs agency co-ordinator. She was put into training as a 'passive drug sniffer' from the age of six months: she sits by where the drugs are, and was also trained to sniff out mobile phones (electronics give off a smell).

"Lots of sniffer dogs retire at around 3 years of age as they become de-sensitised, but Bella was retired because of the arthritis in her paws. We were told she had a touch of arthritis, which was something of an understatement.

"Her joints had been under a lot of strain because she'd have to jump over bunk beds in bursts of frenzied activity for 20 minutes at a time. She was kept at home by the handler in outside kennels, so no playing, or sent to prison cells to do a sweep.

"We were told she could handle two, 40 minute walks a day, but we're not sure they're right about that. She was struggling to get out of her basket so we put her on a large mattress, and cut the walks to once a day.

"She's limping periodically, more often, but not so severely for a few days and then it goes away. She seems most comfortable when she lies flat with her limbs 'splayed out.' She holds up her paw if she doesn't want to go out. Her last diagnosis was a rather general one in that she was said to have fairly extensive arthritic growths, especially

in the toe area and rear leg joints. Bella takes glucosamine and chondroitin, and fish oil, but these don't seem to have worked that effectively. The vet suggested we put her on senior foods, and she has a high-strength flexible joint tablet every day. We also massage her legs.

"The vet said not to throw a ball for her as such a burst of acceleration is not good, and we don't let her clamber or jump over beds. She copes well generally and has found her bark again."

My DOG has arthritis

Breeds affected by inflammatory arthritis

Auto-immune conditions, such as rheumatoid arthritis or lupus, generally affect small and toy dogs of both sexes usually between the ages of two and six years.

Other breeds which can be prone to inflammatory arthritis include West Highland Terriers, Bernese Mountain Dogs, Cocker Spaniels, Boxers, and Beagles. Cross-breeds, such as Labs-Pitbulls, can get it, too.

Doing what comes naturally

When form matches function, arthritis is less likely to occur. Greyhounds, for example, are designed for racing, and rarely get arthritis – or only in old age, if so. However, Border Collies, following their natural herding instincts, can still develop arthritis in hips or elbows, for various reasons.

And Springer Spaniels, a breed developed as a dog-for-all-seasons and every purpose, can suffer from arthritis, perhaps because of too much pressure on joints before they're properly formed.

When Tigger grows up he could earn his keep catching criminals.

Injuries

Although any dog can sustain an injury when involved in even the most basic activities, certain breeds – including Labrador Retrievers, Golden Retrievers, Mastiffs, Boxers, Poodles, and Rottweilers – are prone to cranial cruciate ligament (CCL) tears in their knees, often resulting in degenerative joint disease further down the road. The reasons for this injury are many and varied, and genes play a part.

Limited gene pool

Pure-breeds are thought to be more predisposed to arthritis because of in-breeding. While the human population is 'out-bred,' dogs have a much more restricted gene pool. In general, in-breeding is considered bad for genetic diversity, which can result in dogs who are less fit and more susceptible to inherited diseases.

The Metropolitan Police Service Dog Support Unit runs a breeding programme which has had considerable success in screening out inherited (and potentially arthritic) health problems among the breeds it uses for various duties, including German or Belgian Shepherds and Spaniels.

Estimated breeding values

As part of their ongoing genetic research programme, scientists at the Animal Health Trust, a Suffolk-based veterinary charity, have been developing genetic health tests called Estimated Breeding Values (EBVs) to assist in selection against joint problems such as hip dysplasia. The use of these EBVs could result in a 20 per cent decrease in HD cases.

EBVs are available for breeds such as Labradors, German Shepherds, Rottweilers, Border Collies, English Setters, Tibetan Terriers, and Rhodesian Ridgebacks via the 'Mate Select' service on the Kennel Club's website.

Hip problems in smaller breeds

Although hip problems are more common in larger pure-breds, and rare in toy or miniature dogs, smaller pedigrees can also be susceptible, as Bobby's case history (overleaf) reveals.

Where form matches function – a greyhound doing what she was designed to do. (Courtesy Retired Greyhound Trust)

Bobby, a Bischon Frise owned by Barbara

"Bobby is ten years old, and just over a year ago, whenever I took him for a walk, he was obviously having problems getting up kerbs.

"One night at bedtime he refused to come upstairs where he sleeps in his bed next to me, and instead walked down the hall to his bed in the kitchen.

"This went on all the next day: there was no jumping up, and he was hesitant when having to climb, so I took him to the vet where X-rays showed he was developing osteoarthritis in both hips.

"Hip dysplasia wasn't mentioned but I know Bischons can have HD, and I might check back with the breeder on this who I'm still in touch with.

"Bobby was put on Previcox for two weeks, and then I reduced the painkillers, and have actually only needed to use them three times in the last year for just a couple of days when I sensed the pain was returning.

"The vet said to exercise Bobby little and often, rather than give him one long walk as then he'd get stiff. Out on a walk recently he was charging around without a care in the world. A couple stopped and, to my surprise, asked me how old my 'puppy' was.

"I was delighted!"

Chapter 6

Diet and supplements

It's probably fair to say that dogs are catching up with people in the obesity stakes, as a result of too much of the 'wrong' food – or just too much food, full stop.

Your dog needs to be lean and keen (though not mean) to cope with arthritis, so being overweight won't help at all.

Dieting your dog

An estimated 50 per cent of dogs are obese, and if this applies to yours, it's time to take action. As any vet will tell you, the best favour you can do your dog is to diet him.

Dieting can obviate the need for surgery for some dogs: a good enough argument if ever there was one for helping your dog to shed those extra kilos.

And feeding puppies a leaner diet during their formative years may also reduce the risk of conditions such as hip dysplasia.

Getting help

Most veterinary practices have weight clinics, and some even hold diet parties, where dogs are weighed together in a friendly atmosphere – a bit like Weight

If food is left lying around, the temptation may be simply too great for Barney to resist.

Watchers®. Canine bonding surely has never been so much fun!

One way vets test whether or not a dog is overweight is by checking how easy it is to feel their ribs, so you could keep an eye on your dog's weight

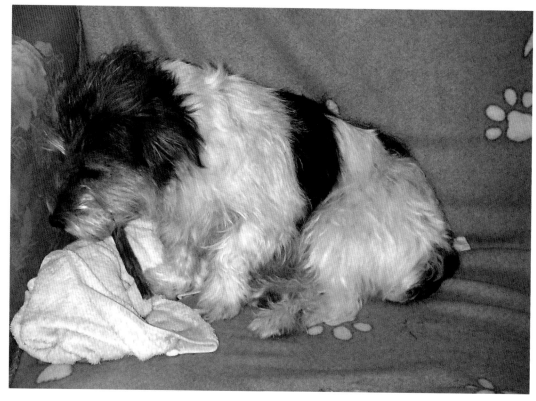

Tilly settles down to enjoy a special venison-stick treat.

this way, or go along to the clinic regularly if you feel happier to let the professionals take charge.

Programmed to eat

Dogs are primarily pack animals, designed to eat as much as they can, as quickly as possible, before another dog gets a look in, so it's best not to leave food lying around the house in full view of your dog, as the temptation to scoff the lot may be just too hard to resist.

And while it might be hard for you to ignore your dog's doe-eyed pleas for treats, aim to limit the times you give in, and try to substitute low-fat treats for high-fat examples. Some dogs like fruit, so you could encourage this

(although never give grapes or raisins as these can be toxic), and others can't get enough of carrots and sprouts apparently, so maybe have a few of these to hand.

Will certain foods help with arthritis?

Whether or not some foods have a positive effect on your dog's arthritis isn't clear, but certainly a healthy, balanced diet can only help your dog to deal with the effects of the disease, and maintain overall fitness.

It can be difficult to know what to give your dog as what's good for us, nutritionally-speaking, might not work as well for them. Your vet – and in some practices, a nutrition nurse – can give you advice about the right diet

Vincent, a young Lurcher-cross being cared for by Ann at Rotherham Dog Rescue

"When Vincent came to us at around ten months old, we noticed he was walking quite awkwardly, with his hindquarters swinging out to one side: he was also very reluctant to run.

"Vincent was diagnosed with severe bilateral hip dysplasia, and I was told the only cure would be a double hip replacement. As this was out of the question, due to the high costs involved, I looked into alternatives that would help Vinnie lead a relatively normal life. I researched different joint supplements, and Mobile Bones (Pooch & Mutt) sounded ideal.

"Subsequently, Vinnie has been castrated, has regular but not too strenuous exercise, and has put on weight and built muscle mass around his hips, all of which – combined with supplements twice a day – has enabled him to have a pain-free life, and also be let off the lead to play and run with my dogs without any detriment to himself.

"Our vet is so pleased with his progress that he's not going to refer him to an orthopaedic specialist unless his condition takes a turn for the worse.

"Vinnie is a smashing young dog who loves everything and everybody: hopefully, one day he'll find his forever home."

Brook, a black Labrador Retriever, owned by Mark & Gerry

"Brook's 11 now, and has been a gun dog most of his life. When he's out on a shoot it's a full-on day with high-pace exercise, and bursts of high activity over long distances. Brook loves chasing pheasants, partridge and ducks, sometimes getting wet and cold into the bargain, which we don't think has helped his arthritis.

"The problems began in his knee and elbow joints, which we noticed particularly the winter before last. During the last season he could hardly get out of bed, and found it difficult to stretch, so he was worked only one day a week instead of three or four, which was a shame as he's a really good gun dog.

"We've been giving him special working-dog food, along with glucosamine sulphate and green-lipped mussels. He has one each of double strength in cheese or meat, and wolfs them down. I also soak his biscuit in water and garlic granules, and he has cod liver oil, too.

"We used to just give him supplements in September, when the shooting season began, but now we give them to him all the time.

"Last year Mark entered Brook for a working test, which he would have won but for a technicality. He's quite sprightly for an old boy; I really think the supplements have made all the difference!"

to follow. Always check with your vet before putting your dog on a special diet as this could cause problems.

A good balance

An under-nourished dog is as unhealthy as an over-fed one, so try to ensure that your dog is getting a balance of protein (in meat or vegetable form); fat for energy; fibre for digestion; essential fatty acids to help control inflammation, and the vitamins and minerals she needs for general good health.

Vitamins such as C and E help to reduce inflammation, and Vitamin D (with calcium) will improve your dog's muscle and bone strength. Minerals such as magnesium – found in salmon and spices such as turmeric – are important, too. (One teaspoon of turmeric daily, sprinkled on food, is around the right amount for a Labrador.)

Fatty acids

Studies on essential fatty acids (such as Omega 3) show they can ease joint pain and help to reduce inflammation and stiffness – though it might take several months for your dog to feel the full benefits. Try to include some oily fish, such as pilchards, tuna or salmon, in your dog's menu. A healthy diet should have a balance of Omega 6 (found in meat such as pork and turkey, or evening primrose oil) and Omega 3.

Supplements

People like their dogs to take supplements because they're not drugs, yet there's no scientific evidence that these actually work. As arthritis symptoms fluctuate, it's hard to tell if the supplements are really helping, or whether your dog is going through a good phase.

Always check with your vet before giving your dog supplements as some may react with prescribed drugs, and can be dangerous if taken in high doses. And ask for advice on which ones to try.

If tempted to give your dog the same supplements that you take, remember that dogs need much higher doses to reap the benefit. Special canine supplements can be expensive, so buying online may help keep costs down.

Choice of supplements

There's a bewildering choice of supplements on the market, and sometimes it's hard to know which to go for, if you want to go down that route. The following are some you could consider:

GLUCOSAMINE, CHONDROITIN AND MSM

Glucosamine is extracted from crab, lobster and shrimp shells, and is believed to play a role in the production and repair of cartilage, while chondroitin sulfate reduces inflammation and supports the glucosamine. MSM (methylsulfonylmethane) is believed to provide compounds that might inhibit pain.

HYALURONIC ACID

This is an essential part of the fluid that lubricates the joints, and plays an important role in keeping joints healthy. As glucosamine first needs to be converted to hyaluronan, this could be a quicker way to achieve the result you're looking for.

LITOZIN

A food supplement derived from wild rose hips, which are high in vitamins C, A, K and B, and contain an essential fatty acid called GOPO that has an anti-inflammatory effect.

DEVIL'S CLAW AND CETYL M

These have been found to have

beneficial effects in reducing the pain, stiffness, and inflammation associated with arthritis, and might also improve mobility. Unlike other supplements which are taken on a long-term basis, a course of CM is usually taken just twice a year.

Benefit for working dogs

Some believe that supplements can improve the mobility of working dogs, and enhance the quality of life for older dogs with arthritis, alongside conventional medicines, a good diet, and weight loss/control.

Natural' might not always be better

Pet food manufacturers have been accused of 'bulking-out' their products with junk. If you have any concerns about feeding your dog commercial dog food you could always check on the individual company's website to see what ingredients it's using, and the amounts.

If you'd rather go for 'natural' brands – some devised by vets – the choice has never been better in supermarkets, pet shops, and online. And if you're happy to cook for your dog, this could be the way to go, as long as the meal is balanced, as dogs need a variety of foods to stay healthy.

And be cautious of the 'wild' diet advocated by some who argue that, as dogs are just wolves deep down, they're designed to eat lots of raw meat and bones. Raw meat has a greater potential for contamination, for example.

And don't forget ... water is the essence of life, helping to keep your dog hydrated and aiding his digestion. Always ensure he has easy access to plenty of clean, fresh water at all times.

Visit Hubble and Hattie on the web: www.hubbleandhattie.com & www.hubbleandhattieblogspot.com · Details of all books · Special offers · Newsletter · New book news **twitter**

47

Chapter 7

Complementary therapies

The choice of complementary therapies increases all the time, or so it seems, and now many have been adapted to suit our canine friends – which, generally, is good news.

These therapies can offer most dogs relief from certain symptoms of arthritis, or some of the side effects of conventional drug treatment. And some owners believe that their dog's life can even by extended with the help of these remedies. Generally speaking, complementary therapies are safe if practised by a qualified therapist from a recognised organisation. A referral from your vet is usually necessary.

Holistic approach

While conventional medicines focus on treating the pain and inflammation that goes hand-in-hand with arthritis, complementary therapies take into account how the whole body is functioning – hence holistic – including lifestyle, personality, and diet.

Complementary therapies are designed to be used alongside conventional medicines and treatments, and not as a replacement, although some owners find their dogs live happily on these therapies alone.

How effective are they?

While scientific research on most complementary therapies is still relatively new and small-scale, early results show that some therapies might indeed help to ease physical (and possibly emotional) symptoms in some dogs.

Each complementary therapy works on a different set of signs, such as stiffness on waking that eases on movement, and while they seem to help in most cases, they can't alter the course of your dog's arthritis or provide a cure, sadly.

The following are some of the more popular complementary therapies available:

ACUPUNCTURE

Based on ancient Chinese beliefs about how energy moves around the body, this therapy uses fine needles inserted through the skin at certain points and left for around 30 minutes, to help relieve pain and promote general well-being. It can be slightly uncomfortable when the needles are inserted, but shouldn't hurt your dog: some even doze off during a session. Acupuncture can only be performed by a vet,

Poppy, a Jack Russell, owned by Paula

"Poppy's now 17 years old, and has been struggling with arthritis for a few years. Sometimes when she tried to get out of her basket she could barely lift her head until she had been moving for a while. I felt the time was near for Poppy to be put to sleep, for her sake, to help her give in perhaps, but we tried her on Carprieve and Mirtazapine and she was okay, just getting by, not too miserable and eating a bit.

"An animal chiropractor then come out to see her, and the next thing Poppy's head was up, her tail was up, and she was walking better, and paced much less in the evenings. This continued for a week-and-a-half. The chiropractor visited again, and there was the same response. Poppy had a good long sleep, then wanted to eat and walk: she skidded into the kitchen at full speed twice after being in the garden, and was generally back to the old Poppy of four years ago.

"After three more sessions Poppy began eating normally again, and was much more jolly and active. Her neck and the top of her spine are now staying straight, whereas before they were reverting to the wrong position. I can't speak highly enough of the treatment.

"This morning we were walking and had to stop and pick up after she'd been to the toilet. I was wearing gloves and dropped the lead whilst I removed them, thinking that Poppy would simply stand and wait. To my surprise, after a moment's thought, off she went and I caught up with her after about 30 metres. A very naughty girl: much more like the old Poppy!

"It's lovely to have her back, even though I know it can only be for a relatively short time."

Maisie, an 11-year-old Jack Russell-cross, owned by Dilys and Nick

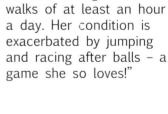

"Maisie has had joint problems, especially with her front legs, from when she was a year old, and the vet said that arthritis could actually help her problem as it would 'roughen-up' the ball and socket joints and make them more stable.

"She's developed stiffness over the years, and does get bouts of pain when she will chew the fur at the tops of her legs, and cry out if you touch her legs, but Rhus Tox 6 works wonders.

"We only dose her when she's stiff or uncomfortable. I give her an acute dose of one tablet every hour for six doses, then one tablet twice a day for about a week, or until she's okay. We've not noticed any side effects. It's also useful for red, swollen and itchy skin, which is made worse by scratching.

"We also believe in lots of regular exercise, and she gets off-lead walks of at least an hour a day. Her condition is exacerbated by jumping and racing after balls – a game she so loves!"

although acupressure (using fingertips only) can be practised on a dog by anyone qualified to do so.

BOWEN TECHNIQUE

This is a 'light-touch' soft tissue therapy named after its innovator, Tom Bowen, and adapted around ten years ago for use on dogs.

The therapist gently moves their fingers over muscle, ligaments and tendons on various parts of a dog's body to promote healing and pain relief.

CHIROPRACTIC

Chiropractors believe that regular adjustment of the joints in the spine and other parts of the body can help keep joints more mobile, reduce pain, and slow further damage.

HOMEOPATHY

Homeopathy is based on the principle of 'like cures like.'

Where conventional medicine aims to suppress symptoms – for example, by using anti-inflammatories to reduce inflammation – homeopathy provokes the body into healing itself.

Homeopathic remedies like Rhus Tox can help dogs with problems such as stiffness that eases on movement.

MASSAGE

During massage the whole of your dog's body is treated in order to reduce soreness, tension, over-compensation, and old muscular injuries.

By manipulating muscles and restoring their function, massage is a natural way of managing your dog's health, prolonging quality of life and promoting well-being. It can help dogs with hip dysplasia in particular.

OSTEOPATHY

The principles of osteopathy are based on the belief that structure governs function, so normal health and mobility

Millie's getting help with the soft tissue problem in her neck, caused by over-enthusiastic play. (Courtesy Tony Nevin, UKSOAP)

will be determined by the structure of a specific joint.

Osteopathy uses gentle manual techniques and soft tissue massage to improve flexibility and range of movement. The treatment will often

Harvey, a Red Setter, owned by Linda

"Harvey the Hound is 14 now, and was so active when he was young, chasing after foxes and crawling under fences. He's had a brilliant life, but has turned me grey!

"When he was nine he was diagnosed with arthritis in his left elbow, and around a year later began showing signs of lameness in his back legs (from the age of six, arthritis was evident in his right hip). We tried various things, including four injections of Cartrophen, and Metacam, which upset his stomach. Our vet then suggested Synoquin, which took a couple of weeks to kick in, but I was concerned about the persistent lameness so I decided to try massage.

"After a couple of sessions I was amazed by how Harvey seemed freer, looser, and was moving more easily; the stiffness had reduced significantly. His quality of life improved – he was able to get up and down more easily, was less stiff, had a sparkle in his eye once more – and wanted to go chasing after bunnies again!

"He showed such good results I decided to carry on regular treatment for Harvey at one session a month: after all, I know how good I feel after a massage so can relate to the benefits for Harvey.

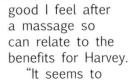

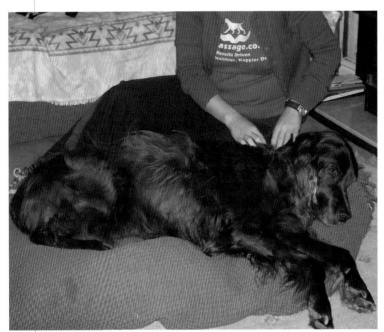

"It seems to have taken years off Harvey – it's given him another four years – and we're able to manage his pain effectively.

"It makes you realise that there are things we can do for our dogs as owners which can help to prolong their life."

Trudy receives some 'hands-on' help.

result in a relaxed dog who's brighter, and more willing and able to exercise.

REIKI

Reiki (pronounced 'ray key') is Japanese for Universal Life Energy. It's a natural, safe, and simple hands-on healing method, with no manipulation. It's believed that Reiki creates and promotes self-healing, balances energies, and revitalizes. Dogs usually become calm and relaxed during treatment.

Chapter 8

Exercise and physiotherapy

Exercise is the pumping mechanism that delivers nutrients to your dog's joints, nourishing cartilage, helping to maintain mobility, preventing stiffness, and strengthening the muscles supporting the joints.

Regular exercise is a good thing for your dog, and is something you can enjoy together – so there are benefits all round.

Ideal exercise

It's never a good idea to over-exercise a very young dog (you wouldn't take a toddler for a five-mile walk, after all), as this can lead to problems later on. Labradors, for example, can develop hip dysplasia if they're over-exercised as puppies and under-exercised as adults. Introduce exercise gradually, and work to a ratio of around 20 minutes' exercise to 40 minutes' sleep or rest while your dog is young.

Little and often

Two or three short (or longer if your dog can tolerate it) walks a day is probably a good guideline for most dogs, rather than one long walk which she sleeps off for the rest of the day, and is then stiff from. If your dog is showing signs of pain, a gentle, on-lead walk the next day should help to ease the stiffness.

High impact

Cartilage, the smooth covering over the end of bones, isn't good at coping with high impact exercise, and could tear. Where possible, try to prevent your dog running up and down stairs, and jumping in and out of the car. Chasing after sticks and balls might seem like fun for your dog, but the twisting, turning action involved puts enormous strain on his hips and knees, and might result in injury which will aggravate arthritis.

Mental exercise

Dogs need mental stimulation, too, otherwise they become bored and, in trying to alleviate this, may misbehave by chewing inappropriate items, or exhibiting other destructive-type behaviour. Joining a local dog club could be one way to provide the mental stimulation she needs, and bonding with your dog doing heel-work to music, or agility classes is rewarding and a lot of fun for you both.

You might also try some of the

puzzle toys on the market that require your dog to work for treats, rather than just being given them.

The benefits of veterinary physiotherapy

Osteoarthritis responds well to physiotherapy, and dogs with auto-immune arthritis can be helped, too, though with more targeted, delicate treatment. Although physiotherapy can't reverse your dog's arthritis, it can relieve some of the symptoms, and with careful management can slow disease progression. Other benefits include minimising pain, and improving muscle condition and overall mobility. It can work well alongside hydrotherapy.

REFERRAL PROCESS

In most cases, a referral is needed from your vet for physiotherapy treatment, as he or she has a clear understanding of your dog's arthritis. As your vet will probably know most of the veterinary physiotherapists in your area, he or she may suggest the one best suited to your dog's condition and requirements.

AN INITIAL ASSESSMENT

The kind of arthritis that your dog is suffering from, the condition of his muscle mass, and general fitness will be assessed at the first session. The physiotherapist will also observe how your dog stands, walks, and sits, and whether (and where) there's stiffness or lameness. If your dog's trying to compensate by shifting weight from one leg to another or leaning to one side, this could indicate problems.

You're likely to be asked where your dog sleeps: on the sofa; on your bed; upstairs or downstairs, and how active your dog is, among other things.

TECHNIQUES THAT CAN HELP

Physiotherapists use a range of techniques, which usually include

massage – to improve circulation and muscle tone, reduce swelling, and prevent adhesions of the tissues following injury – and passive stretching techniques. You can be shown how to do these so that you can continue treatment at home: at least twice a day, ideally.

BEGINNING TREATMENT

Initially, the physiotherapist will work to the principles of PRICE: protection, rest, ice, compression and elevation. Rest protects the affected limb(s) from further damage, and allows time for the inflammation to subside. Ice or cold treatment can constrict blood vessels, helping to reduce blood flow, swelling and inflammation, as well as relieving pain. (In most arthritic cases compression and elevation of a limb don't apply, and rest is considered the best way to protect the limb.)

GRADUALLY INCREASING EXERCISE

After your dog has rested, your physio is likely to recommend a gradual increase in on-lead exercise, normally two or three five-minute walks daily for two weeks, gradually increasing by five to ten minutes a week. This regime gives your dog healing time, and slow, on-lead exercise increases weight-bearing, stride length, and stamina. Off-lead exercise can then be re-introduced.

Dogs with severe arthritis who cannot manage off-lead exercise may become bored with being on the lead all the time, so try to vary their exercise regime. Walking up inclines will help to extend and strengthen their hind limbs, and walking down slopes will help stretch their forelimbs.

TIMESCALE

It will usually take 3-6 months to reduce your dog's joint discomfort, and build muscle mass and general fitness. For the first 2-3 months, you might see

Buddy, a four-year-old Collie-cross, owned by Steph

"Buddy was born in the wilds of Ireland; found at the side of the road with his six brothers and sisters lying next to their dead mother, so he didn't have a great start in life.

"When I collected him from a rescue centre, I noticed that, although he wasn't lame as such, he wasn't weight-bearing on one of his hind legs. This didn't stop him dashing around, though, and he broke his toe. He was diagnosed with hip dysplasia at 9 months of age.

"I was interested in doing agility classes with him as a hobby, but wanted him to have a good quality of life, so I got a referral to canine specialist Dr Shahad Mohammed, at the WitsEnd Rehabilitaion Centre in Leicester.

"Dr Mohammed came up with an exercise programme which has really helped. Buddy's exercise was initially reduced to slow, on-lead walks. After two weeks the amount of time Buddy was walked gradually increased, and within four weeks he seemed to have less discomfort in his hip. Over the next two months more time was spent walking until he was coping with around an hour a day.

"As Buddy was managing well we decided to begin agility on small jumps, and gradually he worked up to full-height jumps and complete circuits (after regular check-ups with Dr Shahad before moving on to each stage).

"Buddy loves agility; I can tell by his body language how excited he is, and watching him jump you wouldn't think he has a problem. He's a keen competitor and came second (out of 15) in a rescue league final, and has qualified for several finals. Within two years Buddy has moved from beginner's agility (Grade 2) to the highest level (Grade 6).

"At home Buddy still has massage, stretching, and a mix of slow walking and off-lead exercise. He takes everything in his stride, and he's loving and well-adjusted – he's a perfect dog, really."

Pepper, an 8-year-old Schnauzer, owned by Victoria

"A year ago, Pepper suffered an FCE, a stroke-like event within her spinal cord, which paralysed her from the shoulder blades back, and reduced her perception of deep pain. She spent nearly three weeks in a specialist veterinary neuro centre, and then had regular hydrotherapy and physiotherapy.

"She now has arthritis in her right leg and drags her feet. Her arthritis was picked up by the physiotherapist, who gave us exercises to do at home with Pepper, including making her step over broom handles (to throw her weight backwards), and weight-shifting and cycling movements for around half an hour a day. The physio also recommended Pepper wear a rubber, waterproof boot which puts pressure on her Achilles heel, forcing her to lift her leg higher.

"To encourage her to walk backwards (which she can't do easily at the moment) we've tried putting her water bowl next to the dishwasher, which is in a small room that she has to reverse out of. We also massage her legs.

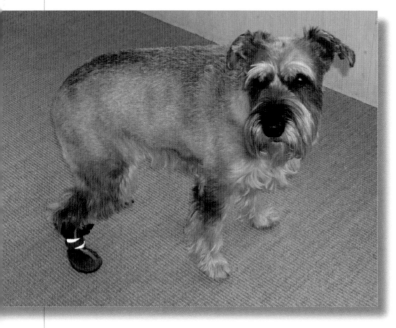

"Pepper is much improved now, and can swim and walk around four miles without struggling. She plays with our other Schnauzer, Oz, who's helped her recovery, and gets her up and about. She bunny hops when she runs, but has a good quality of life.

"After everything she's been through she still lives life to the full."

the physiotherapist weekly, after which visits will probably reduce gradually to once every 3-4 weeks on average.

WORKING TOGETHER

Your physiotherapist should explain fully why your dog has become arthritic, and why each treatment is important at each stage. Feedback from you – is your dog better or is there a flare-up in symptoms, for example – will help ensure that rehabilitation is successful.

HELP WITH DIAGNOSIS

A veterinary physiotherapist should be able to spot the signs of arthritis in dogs referred to them – even if for a different problem – and suggest exercises that will help.

Walking up gentle slopes could help strengthen Monty's hind limbs.

Visit Hubble and Hattie on the web: www.hubbleandhattie.com & www.hubbleandhattieblogspot.com · Details of all books · Special offers · Newsletter · New book news **twitter**

59

Chapter 9

Hydrotherapy

Hydrotherapy is essentially safe, all-round exercise, which takes place in a warm water pool or on an underwater treadmill. The beauty of the treatment is it can help manage your dog's arthritis without putting her poor old joints under any more strain than they already are, and increase mobility at the same time.

And most dogs seem to like it, especially those breeds – Spaniels and Terriers – which can't help throwing themselves into water at the drop of a hat. Those dogs who only initially tolerate hydro, usually come round in the end, especially if bribery, in the shape of treats, is involved.

Whether your dog loves it, or has a take-it-or-leave-it attitude, she can definitely benefit from a few sessions in the hydrotherapy pool.

Land versus water

Each time your dog moves on land a shock wave is transmitted up the limbs and absorbed by bones, tendons and joints.

While weight-bearing exercise helps to maintain healthy, strong bone, if this exercise is severe, repeated too often or is high-impact, it could damage or weaken an arthritic joint, or one recovering from injury or surgery.

With hydrotherapy, the water does the weight-bearing instead, although the muscles still have to work hard: harder even than on land because of the resistance of the water. Water-based exercise uses 30 per cent more oxygen than similar, land-based activity, and dogs often move faster in the pool than they would on land. It's a good, all-round, cardiovascular exercise for your dog, which can only help him to cope better with arthritis.

Other benefits

These include strengthening and toning most muscles; increasing muscle bulk, and relieving joint pain, swelling and stiffness. Chances are your dog will feel more confident back on land, too, which is a bonus.

Reversing muscle wastage

Your dog's muscles can begin to atrophy (waste) just three days after becoming immobile, so it's important to prevent further weakness, and possible injury, by re-building these deteriorating muscles as soon as possible through safe exercise such as hydrotherapy.

Increasing range of movement

As the water takes the load, reducing the weight on a painful joint, or one that's healing, your dog is more likely to stretch his legs further in the water – and might be happier doing this when back on land, too.

Improving circulation

Warm water increases blood circulation to the muscles, helping them to relax. It also boosts the supply of oxygen and nutrients to the tissues, and helps flush away waste products.

Mood enhancer

Apart from the physical benefits, hydrotherapy can have a positive effect on your dog's mood, as well as provide mental stimulation.

In the water, dogs can safely play ball, which they might not be able to do on land, and can be quite perky when they leave the pool: there's often an overall improvement in their mental state within a couple of weeks.

Helping with injuries & joint problems

Hydrotherapy can help with hip and elbow dysplasia – especially with young dogs still on the lead – and recovery from injury, such as cranial cruciate ligament damage.

This kind of exercise can also help combat obesity, alongside weight loss and a healthy diet.

First steps

Hydrotherapy usually requires a referral from your vet to ensure there's no reason why it might not be appropriate for your dog.

A hydrotherapist will give your dog a head-to-tail health check, looking at circulation, checking for any lumps or bumps, and the condition of her ears, before she goes into the pool for the first time. While hydrotherapists can't diagnose problems, they'll refer your dog to a vet if they suspect something's amiss.

Trained therapists

The Canine Hydrotherapy Association represents most of the hydrotherapy centres in the UK, with 60 practitioners on its register – your vet will probably know one in your local area. Most pet insurance companies will only pay out for hydrotherapy treatment practised by a therapist from the CHA.

Time in the pool

This rather depends on why your dog is having treatment. Dogs usually attend up to twice a week, and the time in the pool can be built up gradually, according to ability and condition. Ideally, your dog should be assessed before and after each swim, including temperature checks (dogs suffer heat exhaustion more quickly than people do, so their temperature in the pool will be lower than it would be for a person).

Gradual increase

If your dog's treatment starts with a session of 3 x 1 minute, this might be increased to 4 x 1 minute, and then 4 x 2 minutes. Session time can increase quite quickly for some dogs, although those who are overweight, or who have heart murmurs take it more slowly.

As long as your dog can cope, it's advisable to go twice a week for the first two weeks, as any less often would mean that the benefit would take longer to become apparent. Long-term, once a week should be all your dog needs.

Some pet insurance firms will pay for a maximum of ten sessions only, so do check your policy first, or be prepared to fund more sessions yourself.

Feedback

The therapist will want to know how your dog coped with the first session – was he stiff or tired, for example,

Mara, a 13-year-old Springer Spaniel, owned by Pam

"Mara ruptured her right anterior cruciate ligament at the age of six, and although her symptoms were classic, our vet couldn't understand why her entire back end was so unstable. X-rays soon revealed that not only had she ruptured the cruciate, she also had one of the worst cases of arthritis in her left hip he'd ever seen in a dog her age.

"The right leg had, in fact, been 'carrying' the left, unbeknown to us. Mara had the operation to replace the cruciate, and our vet was also talking about her having a hip replacement for the arthritis. I was reluctant to put her through more major surgery at that time, so took her for hydrotherapy.

"She made marvellous progress and, within weeks, was walking happily again, albeit with a slight limp. After several months her mobility was such that you wouldn't know she'd experienced any problems at all. Any talk of a hip replacement was forgotten as she continued to lead a normal active life, apart from my insistence that she keep up her fortnightly hydrotherapy sessions, which she loves as, being a Springer, she's addicted to water!

"The staff at the hydrotherapy centre told me that, realistically, we'd probably only be looking at a maximum 12 years' active lifespan for Mara, given the severity of her arthritis, but this seemed good to us as she was only seven at the time.

"Fast forward just over six years and Mara is still with us. She happily keeps up with our younger, five-year-old Springer, Bracken (with whom she still enjoys a daily rough and tumble), and still has her fortnightly swims where she's now the pool's longest-attending 'client.'

"Bracken has just begun to show signs of cruciate weakness which I want to treat before it becomes a serious problem."

Teal, an 11-year-old black Labrador, owned by Nicola

"Teal is lame in her left leg, and has arthritis in her elbow, which is secondary to osteochondritis. She had eleven hydrotherapy sessions, which I think helped in her recovery from surgery (to remove a lipoma), and improved her muscle tone.

"Initially, Teal seemed only to tolerate the sessions (with help from some meaty treats), though did appear to get more from the therapy as the sessions went on. She wore a comfy buoyancy collar to keep her head up so her ears didn't get water in them, as she tended to roll in the pool. She liked being hosed down and having her coat dried afterwards.

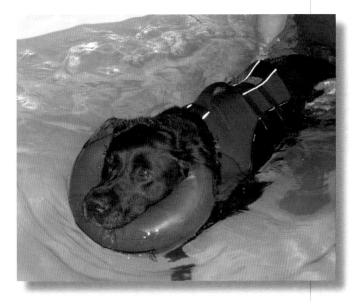

"Presently, she's enjoying her walks, and doesn't want to be left at home to snooze as she did pre-surgery. While she loves food – Labs are the foodies of the dog world – we are strict about how much we give her, as keeping her weight down seems to be a big factor in maintaining her comfort.

"Her mood is much brighter in general, and she seems happier in herself. Should she regress at any point I'd definitely return to the pool as I do believe it was of benefit to her."

perhaps indicating that his spell in the pool was too long or too strenuous. Go by your instincts: if you think something's wrong, tell the therapist and he or she will make adjustments.

Swimming gear

Dogs either wear a buoyancy aid (as much for your peace of mind as anything) or a harness. Some hydrotherapists go into the pool with the animals, others don't. And some also give massage in the pool.

A typical pool

The pool is like a large fish tank. Many practices have underwater treadmills for muscle toning work, though most arthritic dogs use the pool rather than the treadmill. You won't be encouraged to jump into the pool, too, however tempting it might be!

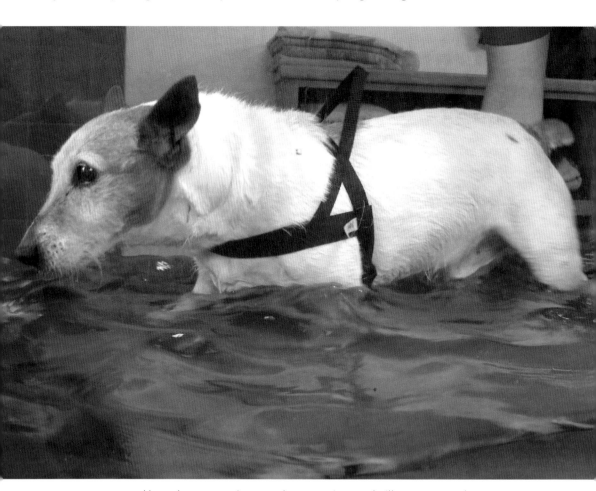

Harry has a session on the aquatic treadmill once a week.
(Courtesy Canine Hydrotherapy Association)

Chapter 10

When euthanasia is the only option

There comes a point in every dog's life when the most loving thing you can do for her is to give her a peaceful and pain-free death. Dogs with severe arthritis, whose quality of life is so compromised that they can't perform even the most basic things without pain, really need to be set free from their suffering.

Deciding to end your dog's life is likely to be one of the hardest and most painful decisions you'll ever have to make.

When is the right time?

There are usually tell-tale signs that let you know when a dog is finding the pain of arthritis more than he can bear. He might be off his food, and even refuse his usual treats. He probably won't be drinking enough water, and could find breathing a bit of a struggle. He'll also probably stay in his basket for most of the day, and lose interest in family life: in essence, he will begin to withdraw. Your vet can help you decide whether the time has come to say goodbye.

Letting go

Most vets agree that dogs usually decide that they want to go, but sometimes hold on because of worry that their owner might not be able to cope when they've gone.

If you think that this might be the case with your faithful friend, you could say something along the lines of: "I can see you're suffering, and I think you'd like to be set free. I'll miss you so much, but I can't bear to watch you in so much pain." Your dog might then feel that with this 'permission' from those he loves the most, he can at last let go.

The euthanasia procedure

Euthanasia is considered a quick and peaceful death (the word in Greek means 'good death'), which usually involves an intravenous injection – an overdose of barbiturate – that typically induces death within 30 seconds.

Previous to this, your vet may give your dog an injection to make her very relaxed and sleepy – almost unconscious – so you can say a final goodbye, before the second injection that stops her heart.

You can stay with your dog during this time, and most owners would want to be there at the end of their

Poppy, a long-haired Chihuahua, owned by Judy & Gordy

"Poppy was 26 years old when she was put down on Wednesday, 13 October 2010, at 11.50am: a date and time I, and Gordy, will never forget. She was one of the oldest dogs in the UK; maybe even *the* oldest at one time, though this couldn't be verified as the only information I had about her date of birth was that it was sometime in September 1984. So, sadly, Poppy never made it into the *Guinness Book of Records.*

"Poppy – then known as Sophie – started life in a high-rise block in London, and the lack of exercise probably didn't help her arthritis. At the age of 12, her owner decided she didn't want her anymore and gave her to some friends, who then decided to move abroad. We fell in love with Sophie the instant we saw her, and joked that as long as she could get through our cat-flap we'd take her! (she could).

"Sophie shivered continually for two days out of fear when she came to us, but then she calmed down, got used to her new home, and never looked back. We re-named her Poppy and she began her new life with a spring in her step, enjoying walks in the country and by the sea, relishing a freedom she hadn't known before. It was a real case of third time lucky for this little dog.

"Poppy was as fit as a flea – really sprightly – and had a great personality. She was very intelligent and had a sixth sense: she always knew what was going on. She was very friendly and polite and would always go up to people to say hello (if they rejected her, she never forgot, and would ignore them if she saw them again).

"She was also a bit of a bossy boots, taking on bigger dogs and cats, totally in charge. She'd never run for a stick, but would

'prance' after it instead. She adored her daily walk, which, if she hadn't had by 3pm, she'd demand in her own special way.

"She lost the sight in her right eye five years ago when she was 21 after a fight with a couple of cats, but it never bothered her or stopped her doing what she wanted to. I think she could have lived until she was 30, or even 35, were it not for the arthritis in her back legs.

"I knew Poppy was giving up when she couldn't jump up onto the low sofa for a cuddle – her favourite thing to do. Her walking became slower, and she was clearly struggling. When on all four legs, she'd half-crouch. She had no quality of life and began to walk less and less, trailing behind me, whereas before she would have been prancing about ahead of me.

"She wouldn't come into the sitting room for ages, until the last day she came in and stood on the carpet and urinated in front of us, as if to say: 'I've had enough.' It was very deliberate. She'd never done that before and would always bark by the door if she wanted to go to the loo, keeping it up until she was let out, even if it was the middle of the night. (I still hear 'phantom' barks – Poppy's bark.)

"I rang the vet who told me to bring her in: it was time for her to be put to sleep and out of her misery. We were privileged to have had Poppy in our lives – we loved her dearly – but over the last two or three weeks had to face up to the fact that we were being selfish by keeping her going.

"We went with her to the vet, and it was all very quick with a massive dose of anaesthetic sending Poppy on her way. We brought her back home; dug a hole in the front garden, and buried Poppy wrapped in her favourite blanket.

"I cried for three days, and for weeks we were convinced she wasn't dead: hoping, somehow, that she might still be alive. That feeling lasted for about a month (we were reluctant to bury Poppy until we were convinced she really was dead).

"Although we've no intention of moving, if we ever did, we'll attach a note to her grave, telling the new owners about lovely Poppy who brought so much joy into our lives."

much-loved dog's life, to comfort and reassure her.

A final resting place
Try and give some thought beforehand to what you would like to happen to your dog's body. This is a very personal decision, and only yours to make, so don't worry about what others say or think.

You can bury your dog in your own garden (check with your local authority), although don't wrap her body in plastic of any kind, as this will interfere with the natural decaying process. Instead, you could wrap her in a towel or her favourite blanket. Bear in mind, though, that if you move house, you would leave behind your dog's last resting place.

If you have had her body cremated, you could scatter her ashes in the garden; perhaps where she liked to lie in the sun or under the shade of a favourite tree. Or you could keep the ashes in a special container in the house, or bury this in a special place in the garden.

Courtesy Dignity Pet Crematorium.

Whilst a pet cemetery isn't as personal – and is a more expensive option – all arrangements are taken care of for you, and you would be able to visit the grave, even if you move house.

Coping with your loss

It's not easy to cope with the loss of your canine friend, and there's no right or wrong way to grieve. Give yourself time to mourn, and don't rebuke yourself for being upset about your dog's death. Only you know what she meant to you, and how long you will need to grieve: there's really no time limit. Cry if you want to – for as long as you want to.

It'll help to remember the good times you had with your dog, and to remind yourself that she is no longer suffering. There are various ways you can remember your dog, and thank her for all the pleasure she brought to your life. For example, the London Veterinary Clinic has an 'In Memory of' page on its website where you can place a photo and special message for your dog, and light a virtual candle for her.

Arnold the Doberman, a rescue dog, nearly 11 years old, and owned by Pam

"We've had Arnold since he was a 12-week-old pup. For the first four or five years of his life, he was the supreme athlete but always had a strange gait: instead of moving his hocks under him like pistons, he would swing his legs from side to side. Turned out, he had osteoarthritis in his legs, but this didn't stop him jumping and racing around – doing what we called 'helicopters.' We used to joke that if he were a horse, he'd compete in the Grand National as he used to take flying leaps at any obstacle put up for him. He had a good life and was a lovely companion and always loved having a game.

"Then, one Monday morning, I came down to find Arnold hobbling on his near hind, and knew from the look on his face that he was in real trouble. I called the vet who administered Tramadol and an anti-inflammatory injection. She thought it was probably infectious arthritis, and she could still save him. He'd been on Metacam and Cosequin, and the combination had made him sick, so I don't know if that played a part.

"I knew I was losing him – you know your animals after a while. It all happened so quickly, and was a shock, but you have to be philosophical. You have them for a while and then you lose them, but I hope I gave him all I could.

"Arnold was put to sleep in his own bed – the vet came out to us so it would be much kinder for Arnold – and he curled up next to me and just fell asleep.

"He was cremated, and I've decided not to bury him in the garden. Instead, when the time comes for me to be cremated, I'll take Arnold back to Wiltshire with me and we can walk the Downs together – if that doesn't sound completely potty!"

Conclusion

Being told that your dog has arthritis is not something you ever want to hear. It's generally a debilitating disease which can take some adjusting to – for you and your dog – and if your dog has the inflammatory form of the disease, it could mean frequent visits to the vet and life-long treatment to manage the symptoms.

On a positive note, however, there's a lot that can be done to relieve your dog's pain and stiffness, including administration of the right drugs, physiotherapy, hydrotherapy, and complementary therapies. And putting your dog on a diet, however hard it might be for you both to stomach, can relieve symptoms and help protect your dog's joints – which need all the help they can get, after all.

Hopefully, the case histories in this book will have demonstrated that a diagnosis of arthritis doesn't have to be the end of your dog's world – or yours. You'll still be able to do all the things you love doing together and, by making some adjustments – such as not too much exercise, particularly of the high-impact kind, while still keeping on

the move – you'll be taking steps in the right direction.

Other ways to help manage the condition for the long term should include making sure your dog eats a healthy diet, with all the vitamins and minerals he needs, and visiting your vet if you see symptoms worsening, or you want to try another drug, perhaps with fewer side effects.

Conventional treatment options might seem limited at the moment, and are largely centred on managing the pain of arthritis rather than offering a cure. But, as we've learned, strides have been made with stem cell therapy and osteoarthritis. And research is under way into the role genes play in the disease, with the realistic prospect of developing an animal model of gene therapy for human arthritis in the not-too-distant future: yet another reason to say a big thank you to our canine friends.

In the meantime, don't despair about an arthritis diagnosis: there's every reason to believe there's life (and a happy one at that) in the old dog yet, as the saying goes.

Useful contacts and further reading

Veterinary/complementary therapy

The Royal Veterinary College (University of London)
Tel: 01707 666333
www.rvc.ac.uk

University of Liverpool Small Animal Teaching Hospital
Tel: 0151 795 6100
www.liv.ac.uk/sath

University of Glasgow Small Animal Hospital
Tel: 0141 330 5848
www.gla.ac.uk

The London Vet Clinic
Tel: 0207 723 2068
www.londonvetclinic.co.uk

Fitzpatrick Referrals
Tel: 01483 423761
www.fitzpatrickreferrals.co.uk

Dick White Referrals
Tel: 01638 572 012
www.dickwhitereferrals.com

WitsEnd Rehabilitaion Centre, Leicester, LE2 8AL
www.info@witsend4pets.co.uk

British Association of Homeopathic Veterinary Surgeons
www.bahvs.com

National Board of Certification for Animal Acupressure and Massage
www.nbcaam.org

European Guild of Canine Bowen Therapists
www.caninebowentechnique.com

Referrals in Holistic Veterinary Medicine
www.holisticvet.co.uk

Canine Reiki
www.reiki4dogs.co.uk

The Society of Osteopaths in Animal Practice
www.uksoap.org.uk

Canine Massage Therapy Centre
www.k9massage.co.uk

McTimoney Chiropractic
www.mctimoney-chiropractic.org

Canine Hydrotherapy Association
www.canine-hydrotherapy.org

Suffolk Canine Country Club
Tel: 01284 811066
www.suffolkcaninecountryclub.co.uk

Trusts, clubs & charities

Animal Health Trust
A veterinary charity dedicated to the improvement of the health and welfare of companion animals
Tel: 01638 751000
www.aht.org.uk

Petsavers (the charitable division of British Small Animal Veterinary Association)
Dedicated to improving the understanding and treatment of diseases affecting our pets
Tel: 01452 726723
www.petsavers.org.uk

The Kennel Club
The UK's largest organisation dedicated to the health and welfare of dogs
Tel: 0844 4633980
www.thekennelclub.org.uk

PDSA
A veterinary charity caring for more than 350,000 pet patients belonging to people in need
Tel: 0800 731 2502
www.pdsa.org.uk

Companion Dog Club
Open to people of all ages to spend time with your dog and meet others
www.companiondogclub.org.uk

National Animal Welfare Trust
A charity aiming to positively change the lives of both people and pets
Tel: 0208 950 0177
www.nawt.org.uk

Pets as Therapy
Volunteers with registered PAT dogs visit people in hospitals, homes and schools UK-wide

Tel: 01844 345445
www.petsastherapy.org

Online sites for prescription treatments; supplements, vet and pet products

www.vet-medic.com
www.animeddirect.co.uk
www.Vetuk.co.uk
www.furrypharm.co.uk
www.dogaids.com

Health food, supplements & natural remedies

Pooch & Mutt
Tel: 0844 247 2122
www.poochandmutt.com

PetSpec
Tel: 01845 565030
www.petspec.co.uk

Pet Elements
Tel: 0845 6432021
www.petelements.co.uk

K9 Natural
Tel: 0115 982 3900
www.k9naturalfood.co.uk

Vet's Kitchen
Tel: 01285 711151
www.vetskitchen.co.uk

Canine Health Concern
Tel: 01835 830273
www.canine-health-concern.org.uk

Health For Animals
Tel: 01925 267818
www.healthforanimals.co.uk

Crematoriums

Dignity Pet Crematorium
Tel: 01252 844572
www.dignitypetcrem.co.uk

Further reading

Books

Complete dog massage manual – Gentle Dog Care
Hubble & Hattie ISBN 9781845843229

Dog-friendly gardening – Creating a safe haven for you and your dog
Hubble & Hattie ISBN 9781845844103

Exercising your puppy: a gentle and natural approach – Gentle Dog Care
Hubble & Hattie ISBN 9781845843571

Liviing with an older dog – Gentle Dog Care
Hubble & Hattie ISBN 9781845843359

My dog has cruciate ligament injury – but lives life to the full!
Hubble & Hattie ISBN 9781845843830

My dog has hip dysplasia – but lives life to the full!
Hubble & Hattie ISBN 9781845843823

Swim to recovery: canine hydrotherapy healing – Gentle Dog Care
Hubble & Hattie ISBN 9781845843410

Dieting with my dog
Hubble & Hattie ISBN 9781845844066

Magazines

Dogs Monthly
Tel: 0845 094 8958
www.dogsmonthly.co.uk

Dogs Today
Tel: 01276 858880
www.dogstodaymagazine.co.uk

Your Dog
Tel: 01476 859830
www.yourdog.co.uk

Forums & social networking sites
www.k9united.com
www.i-love-dogs.com/forums
www.petforums.co.uk
www.labradorforums.co.uk
www.forum.hmedicine.com

Annual events
Crufts (Kennel Club event at NEC Birmingham)
www.crufts.org.uk

Discover Dogs Show (Kennel Club event in London)
www.discoverdogs.org.uk

All About Dogs (Aztec Events)
Shows in Newbury, Suffolk & Norfolk
www.allaboutdogsshow.co.uk

All of the foregoing information was correct at time of going to press and no responsibility is taken for omission or error. Inclusion does not confer endorsement by the author or publisher.

Visit Hubble and Hattie on the web: www.hubbleandhattie.com & www.hubbleandhattieblogspot.com · Details of all books · Special offers · Newsletter · New book news **twitter**

73

swim to recovery
canine hydrotherapy healing

Hubble & Hattie

Emily Wong

Gentle Dog Care

ISBN: 9781845843410 £12.99*

"... illustrated with clearly drawn illustrations and radiographs ...
useful to dog owners to improve their understanding of the benefits
of hydrotherapy ... may be most beneficial to owners to help them
understand how hydrotherapy fits into a rehabilitation
programme" – *Veterinary Record*

the complete dog massage manual

Julia Robertson

Hubble & Hattie

Gentle Dog Care

ISBN: 9781845843229 £12.99*

"Clear photos and technique descriptions to help you massage your
dog safely and effectively" – *Pet Magazine*

living with an older dog

David Alderton & Derek Hall

Gentle Dog Care

ISBN: 9781845843359 £12.99*

ISBN: 9781845843571 £12.99*

exercising your puppy
a gentle & natural approach

Julia Robertson &
Elisabeth Pope

Gentle Dog Care

Index